Digital Wonder

Choosing the Right Path Amid Constant Change, Endless Choice, and Digital Living

JULIAN BERGQUIST

DEDICATION

To Generation Y, the Digital Natives, the Emerging
Leaders of our Age.

May your potential be fully realized as you tackle the big
problems we face today, re-orienting our society to what
matters most.

"The most beautiful thing we can experience is the mysterious. It is the source of all true art and science. He to whom the emotion is a stranger, who can no longer pause to wonder and stand wrapped in awe, is as good as dead — his eyes are closed."

- Albert Einstein

Contents

PREFACE   1

ONE   9

TWO   11

THREE   13

FOUR   15

FIVE   17

SIX   21

SEVEN   23

EIGHT   27

NINE   29

TEN   33

ELEVEN   35

TWELVE   37

THIRTEEN   39

FOURTEEN   41

FIFTEEN   45

SIXTEEN   49

SEVENTEEN   53

EPILOGUE   55

"It took millions of years for man's instincts to develop. It will take millions more for them to even vary. It is fashionable to talk about changing man. A communicator must be concerned with unchanging man, with his obsessive drive to survive, to be admired, to succeed, to love, to take care of his own."

- Bill Bernbach

PREFACE

The Prophet lies on his death bed. He phases in and then out of consciousness. In his conscious moments, he speaks pithy words drawing upon his years of experience. In his lifetime he has seen the transition from conventional living to digital living, the quickening of day-to-day change, and the deluge of choice resulting in decision fatigue.

He puts his vision and wisdom into short conversations. Each time he rises from the edge of consciousness, he speaks clearly and plainly to the emerging leaders who have come to honor him and pay their respects. He then retreats into his own world for a few more minutes that he might find some pearl of immutable wisdom. Repeatedly he rises from peace back into the struggle of life, in the hope that nothing that might help these future leaders will be left out.

I have the great fortune to make the Prophet's final words available to you, the technologists, the entrepreneurs, the future leaders of the world. His words are spoken to you so that you may discover your path, and pursue your life's purpose.

I believe the Prophet's words give us a chance to step back and be amazed again or for the first time – amazed at what we have created with our minds, our creativity, and our collective persistence.

These utterances, recorded below, are the Prophet's final gift of love. After speaking these words, he breathed his last.

A World of Constant Change, Endless Choice, and Digital Living

We have completely transformed the way we create and consume knowledge. What was once performed in theater or recited as chant, was replaced by printed books. Then came phones, radio, and TV, followed by the deluge of digital technologies that allowed us to create in ways that were completely digital, separate from the analog world.

More than anything else, the speed at which thoughts are now communicated across distances has increased. This instant communication has expanded the scope and scale of thought exchange more than the speed of the communication; it was always possible to exchange such ideas face to face by talking – but not with someone living on the other side of the world.

The technological capabilities boggle the mind, even for the people who invent them. The possibilities and consequences are seldom foreseen. Human beings are creatures that discover, envision, and create. The thirst and craving to explore and create runs in the deepest channels of our being. At this point in our growth, our ability to love, experience and create is the realm in which we seek to live.

We are in a world where we are instantly in touch with anybody who has agreed to receive our communication. We can now instantly share our digital creations.

Unfortunately, we mostly waste the right-in-front-of-us opportunity to do something great. Instead we banter back and forth on social media about the basics of our day-to-day existence: making money or attending to some aspect of our daily obligations, or showing off a picture or idea that expresses us in a favorable light.

Now that we have converted to digital, the sentiment

of many early-adopters is that something was lost. A certain quality of connection with each other was lost.

In a technological world where we expect instant gratification and communication, we have opportunities and pathways in every direction we look. Yet we are disconnected while seemingly connected. The question is: how do we choose the right path to pursue our dreams?

Bringing Wonder to All that We Do

In alignment with the words of the Prophet, I assert that what has been lost is wonder.

We became desensitized to the pure amazement that is our birthright as human beings. At the rate at which technology advances life and accelerates change, are we still connected with that sense of wonder that has made all of this possible?

One could stand in awe of who we are as human beings, at what we are capable of, and how far we have advanced. We've accelerated in all the sciences, with digital living as a backbone.

So are we human or are we more like gods? Surely we would be seen as gods even viewed from a few decades ago. One can almost view the digital world in the realm of the sublime.

Yet I stand in awe of something else; not who we are as humans, but rather what this experience points to.

It seems to be surging forward in a direction that very few of us question or even stop to ask the question: "To what end?"

Do you have a sense that there must be more than this? Great scientists, when pondering the source of their invention and creations, point to a world just outside this in which we live. Their sense of wonder enabled them to touch the deeper truths.

Wonder goes beyond curiosity for knowledge; it is amazement at our very ability to know.

The Prophet, through his words, teaches us to bring a sense of wonder to everything we do, such that we discover new ways of thinking, acting, and accomplishing.

By bringing wonder to all that you do, you can invent the life of your choosing and create your life's purpose. This includes launching your ideas into a movement that makes a profound difference for people and the things they most deeply care about.

I'm inviting all technologists, scientists, manufacturers, and providers of services to take a look at where we are taking this planet, and why? Bring wonder to your life, so you may discover and achieve your full potential and make a larger contribution.

Julian Bergquist
Technologist, Entrepreneur, and Business Architect
San Diego, CA
February 2015

"Your work is going to fill a large part of your life, and the only way to be truly satisfied is to do what you believe is great work. And the only way to do great work is to love what you do. If you haven't found it yet, keep looking. Don't settle. As with all matters of the heart, you'll know when you find it."

- Steve Jobs

ONE

With Minutes to Live

My time has come.

I love you. I care about your life, and about what happens to your generation.

I'm concerned that we are, once again, losing our way, and it is, once again, the task of the young to set us on the true path, as marked out by our radically changed circumstances. We need your youth, vitality and fresh thinking to chart a new way.

I once had the essential truth when I was your age. The funny thing is, I lost it. I didn't believe that I could possibly know the key to being happy when everywhere I turned experts and gurus offered something else to consider.

So I got lost. I knew, but I forgot. What made me forget? Struggle. I lost myself trying to make my way in the world. I "grew up," which meant immersing myself in convention and letting go of child-like wonder, and in growing up I grew stale. As we all are, I was faced with many challenges, and I confused success in meeting these challenges (making money, raising a family, achieving career success), with happiness. I achieved success, but happiness remained elusive.

And so, just as many others who find their successes unfulfilling, I went on a quest, an inner journey of growth and development. I met with hundreds of gurus and renowned masters in their fields. I attended hundreds of lectures and gatherings on leadership, personal mastery, health, finance, and business. I invested tens of thousands of hours studying, practicing, and mentoring others.

I don't regret that time invested, don't get me wrong. But it did not let me discover the key to happiness. I didn't need to learn it from someone else or to discover it, because I knew it from the beginning. I had just forgotten.

Here is the key to living a life full of joy, love, and happiness, which every child knows and most adults forget. Bring a sense of wonder and awe to whatever you are doing or whatever challenge you face. You can then be happy and find true success.

TWO

A Sense of Wonder

Happiness comes from within, from a sense of wonder and awe that every child has and most adults, alas, have lost.

Knowledge also comes from that same sense of awe and wonder. From this limitless fountain comes an awareness that there is more to know, more to create, and more to be. Another word for it is curiosity. All children have it. Most adults have lost it.

Not only is that sense of awe and wonder the key to happiness, it is also essential in the creation of all arts and knowledge aimed at solving lesser, more immediate problems. Without it, only rote-learned, tried-and-true solutions can be applied, and we, as human beings, are born for more than that. The great advances of our past, which have built the amazing society in which we live today, were all accomplished by men and women who stood wide-eyed in wonder at the world, who sought to understand it in sheer amazement.

These men and women represented only a tiny leavening in the great mass of humanity. Imagine what could be done if that sense of awe and wonder were the norm instead of the exception.

What if each person could contribute their full potential towards a fulfilling future for all of us?

Here is the answer: wherever there is a problem, something wrong that needs to be solved, bring a sense of wonder.

THREE

The Source of Happiness

Only you can make you happy.

You have already learned or will learn many techniques or approaches to handling most situations in life. In the end, however valuable these techniques may be for solving immediate problems, they are useless for solving the bigger problem, giving you access to what matters most and bringing happiness.

We project our happiness out into the world. It's natural to do this, because when we desire something, gaining what we desire gives us a momentary satisfaction. We think, "getting this — a new job, a new toy, a new lover — made me happy." So we pursue more and more of whatever we associate happiness with, thinking that will make us truly happy. But it never does!

The source of happiness is not in having what you want. The depths of human wants and desires are insatiable and endless. Marketers know this, they appeal to it and sell to it. But marketers couldn't fool us in this way if we were not already fooling ourselves.

Happiness isn't something you acquire, or that is given to you from the outside. It's a baseline state, burning inside like the sun, illuminating your life. It comes from within, and is with you through all the striving, whether you get what you want or fail to do so.

Happy people still strive, because happiness doesn't extinguish desire, but the striving doesn't define them and they don't confuse the satisfaction of desire with happiness.

More money, more sex, more love, or more possessions may give you a momentary satisfaction. But only you can make you happy.

FOUR

Engage the Depths, Not Just the Surface

Some may suggest that the key to discovery and learning is asking the right questions. Though the journey of self-discovery ends in knowing yourself, it doesn't help you reach your full potential. It only engages the surface of the mind.

By engaging the depths as well as the surface, the heart as well as the mind, you can experience and expand yourself, not just intellectually know yourself. You can unlock your greatness, enabling you to continue to grow and expand.

In fact, "knowing yourself" can even be counterproductive, because all such knowledge involves limits: each idea of what you are is also an idea of what you are not. You must get such ideas, such thoughts of yourself, out of the way, in order to see, experience, and understand who you are, which is different and deeper than mere knowledge. Self-"knowledge" often takes the form of limited thinking, beliefs, and perception that cloud real seeing. "Knowledge" of yourself will often keep you small.

The same is true of knowledge of anything else. For technical purposes, as a matter of utility, such limited knowledge has its uses and can be very important, but for real understanding it's important to engage the depths as well as the surface.

Remember: whatever you discovered previously about yourself is not who you are today. So keep going deeper, forget what you discovered about yourself before and explore who you are today, rather than who you were yesterday.

FIVE

Waiting for Experience Often Amounts to an Excuse Not to Plunge in and Try

What is your vision of what you can be? What is your life trajectory? What is a real expression of your hopes, dreams and desires for a life well-lived?

Why aren't you doing that now?

Perhaps you are waiting to be ready, to have more experience. You think you require more education or training, or to spend more time working on what you are doing. You figure you need to get a few more jobs under your belt, and build a more impressive resume – then you'll go for it!

You do all of this basically because you lack the confidence to attempt your desire in the present moment, instead deferring the attempt to a future that never seems to come.

Of course, some training and experience are necessary to develop the skills to be effective on your journey. But for most people, what's lacking isn't knowledge. It's courage. There's a difference between gaining the skills you need to succeed, and using the process of gaining those skills as an excuse not to try right now.

The difference is mostly one of mind-set and attitude. If you are emotionally ready to succeed, you will chafe at the bit. You may know that you need to learn more in order to succeed, but you will want to put your new learning into practice as soon as possible – preferably while you are still learning it.

That is the way to live since you will never stop learning. Which means that learning and gaining experience aren't things you do before you go for your dreams, but things you do in the process of going for what matters most to you.

In contrast, if you feel completely confident that you can succeed, that's a sign you're doing it wrong and being too cautious. The pursuit of anything truly worthwhile requires a level of uncertainty and discomfort every single moment. Step out into the unknown and go for it.

Your fear that you don't have what it takes isn't real. It's a shell built around yourself that keeps you small. You can cultivate a sense of trust and confidence by bringing curiosity and wonder to your view and pursue your dream now.

JULIAN BERGQUIST

SIX

Create Each Moment as a New Act, Not Dictated or Determined By the Past

Don't create your life based on your past. Create each moment fresh as a truly creative act.

The world will tell you that your resume is who you are – That your education, experience, and accomplishments from the past will accurately predict what you are now capable of.

Those things do provide some hints of your likely performance in the future, or capability in an area, but not enough to really know who you are, from this point in time, facing forward, or, even more importantly, what you will become.

You think employers, investors, and business people make their decisions about you from your resume (your past). What they really long to see is who you are and whether they can count on you. That can only be seen in how you approach the moment with them now.

Life takes courage. Your past experience is not the

gauge for creating and evaluating opportunity now. It is predictable and boring and leads to a small life, assuring you that your full potential will remain unexpressed.

Instead, stand there and declare what you are going to do. Then go fill the gap between the present and what you've declared will now be.

This is true leadership.

SEVEN

More Options are Not the Key to Happiness

Endless options seem great and liberating, but they amount only to new ways to scratch an itch. This is not the key to happiness. Consider the vast volume of options now available in every category. With technology, we develop and deploy digital solutions almost instantly and for free. 3D printers allow for creating almost every type of physical object you can imagine. New technology enables us to be perfectly matched for the preferences we seek, having our own digital butler available at our beck and call.

The implicit assumption is that finding what you want will provide some sort of happiness and fulfillment.

I promise, it won't. Success is actually built on prolonging the fulfillment of immediate desire.

The source of happiness is not in having what you currently want. There is a certain level of nobleness and satisfaction in giving people what they are currently asking for, that tells us we've been of service to humanity and the world.

But is giving people what they want contributing to humanity reaching its full potential? When the fulfillment of desires does little more than appease the reptilian brain?

Look for yourself.

Each generation launches with its own set of restrictions and unique hopes for what it can become, traditionally with hopes for a leap above what its parent's generation did. Your generation now seeks to create endless options, with ultimately every individual a creator, contributing to an endless set of choices.

So looking at and thinking about the many options you have, remember that you are actually creating the options. Nobody is to blame but you. Though my generation laid the technological foundation, you decide what happens next.

You must begin looking at what you are creating and why you create it.

We are now creating a massive set of options from which you now have to select. It is your desire to create variation to what is currently available.

If the volume gets too much to bear, you control the throttle.

But please keep in mind that the options will not buy your generation happiness.

So bring a sense of wonder to the mess we've made. To what end are you now creating the options and variations. Why?

What does it provide but a false sense of a deeper need having been met, when it was just scratching another itch?

Find a new context for viewing this digital life that gives this growing selection of options. If you can create that new context, you can be the messiah for your generation.

JULIAN BERGQUIST

EIGHT

The Most Amazing Thing, the Source of the Greatest Wonder and Awe, is that Anything Happens at All

With our minds, we have built the most wonderful structures of theory to explain how the world works. We can model the tiniest subatomic particles and the biggest galactic clusters. We can describe the fields that govern space and time, the emergence of life from the seas, the dance of the synapses in our brains.

Our knowledge keeps growing exponentially and we have no idea what the end will be. But through all of this, there is one mystery that remains impenetrable and always a source of wonder and awe, and that is the fact that anything happens at all.

Why are you here? Why does the impact of light upon your retinas and the processing of that energy by your optic nerve and your brain, give rise to the perception of color, of shape, of beauty or even ugliness? We can see how the information leads to responsive behavior, but that isn't the question. Why are you here, watching it all happen? And why is it, or anything else, happening at all?

Those who have lost the sense of wonder and awe and trudge through life with their eyes to the ground, more than anything else, have forgotten that consciousness and existence itself is strange and inexplicable, and something about which to be amazed.

NINE

Choosing the Thing that Matters Most in Any Situation and Pursuing that is the Key to Achieving Success

We have so many options before us today, and so many possible goals to pursue, that we lose track; we spread our efforts out too much and dilute those efforts, and fail to accomplish much of anything.

Complicating even that situation, is that we confuse achieving our goals with the "pursuit of happiness," an unfortunate phrase from Mr. Jefferson that suggests happiness is out there and we are capable of finding it by acquiring more money, more things, or more life experiences. Happiness cannot be pursued. You already have it, and are the source of it yourself.

But many other things can be pursued and are worth pursuing, and this is where the dilution of effort and the great confusion of our times comes into play.

What is the purpose of your life? What is it that you are here to do? Is it different for each person? No, in general terms it can be described the same for everyone. Be alive, do stuff, learn, and grow. That is your purpose for being alive.

Simple, isn't it? But while you can hardly help pursuing your purpose, happiness is found in your own core and just needs to be awakened. Success and achievement in your more immediate goals follows a slightly different set of rules.

There's a simple decision for you to make as you confront the dizzying array of options before you and various goals that you are able to pursue. Choose the one thing that matters most among them. Pursue that first and foremost, if necessary at the expense of lesser goals. You can have more than one goal in front of you at a time, but always keep that one thing on the front burner and at the top of your list.

What is that one thing? That is something for you to decide. What I want to tell you is not what it is, but that it is a choice you need to make.

For example, once you begin a project you may find an earlier idea you put aside creeping back in or you might see your original plan morphing into something else. Though the shift may be an enhancement, it is more likely a conflicting intention or commitment. You can't have two masters at the same time. Take note if the shifts are moving you and the project closer to completion of the original vision or confusing it. There is nothing wrong with having more than one intention or commitment, just not at the same moment in time. Otherwise your goals will compete with each other and produce a paralyzed mind.

You will have no idea about the outcome of a decision you make now. You can't be sure what new opportunities will open up, and what other ones will be closed by you making a specific choice. You can only know the outcome by looking back.

So when you look back at what you've accomplished, or not accomplished, what will you gauge your success on?

In the end, success isn't so much about achieving goals as it is about learning and growing. Look back with appreciation, with a sense of wonder about the learning and growth you have had. Realize that this was your journey and you were in control the entire time. Choose from there how you will navigate the next phase of your life.

What matters most is to "will one thing," or "the Good" as Kierkegaard says. What we know is that making a difference and having experienced love is what matters most. What does that mean exactly? This is for you to discover and decide for yourself.

TEN

You are Not Your Feelings

You are a bag of chemicals. Your body is the most amazing pharmacy, making and deploying just the right chemicals for the right time. The adrenalin rushes, the highs and the lows, the sex, passion and intimacy, and all the emotional dramas release rushes of chemicals for us to experience and often enjoy. And then we become addicted to the chemicals. We enjoy them, we dwell on them, and we seek experiences that release them. Often we engage in behaviors that don't give us what we want most in life.

You are not your feelings even though it may seem like it. Don't let them run your life. Choose your own course regardless of how you feel in the moment by focusing your behavior on things that support your end goals.

Bring your curiosity to this: that your feelings are part of your experience of being alive, that you crave them, and that you are addicted to them.

Once you get how attached you are to your feelings, just let the entire thing go. You can choose to live powerfully. You will feel many things in life no matter what path you take. Even though you desire to experience life, that doesn't mean you need to let the chemicals call the shots. You call the shots, and choose the direction of your life. Whatever you choose, there are plenty of opportunities to experience and find ways to stimulate the chemical factory. Living life on the edge will give you the experience of joy and aliveness as you push the edge of that which you are capable.

ELEVEN

The Source of Invention is the Ability to Put Aside What You Know and Wipe the Slate Clean

The source of invention is your ability to put aside what you know, and your assumptions, and start fresh, to make room for the wholly new. Each day challenges what you know, not by showing that it is wrong, but instead showing that there is another level or something even more advanced and truer than what you have now.

The key is putting all that you know aside, wiping the slate clean. I cannot stress the importance of this enough. Move everything off your desk. Bring out a new sheet of paper. Start a brand new file. Literally make the space of nothing. Now go deeper, or, from a new angle look for something that is not familiar.

This makes room for something new, something unexpected, that you're not seeing now. Something magnificent.

This may be difficult to pull off when your old thinking is still inside you, deep in your memory.

Example: next time you think you have a brilliant thought, ask yourself "Is this familiar?" You'll likely notice that you have had this thought before, or something close to it. You're mostly regurgitating what you have previously thought, learned, or done before.

Challenge your originality, challenge your notions of your own brilliance. There is little new under the sun. Every now and then again somebody actually "invents."

Yet if you can go to the edge of wonder, the ineffable world where thoughts arise, you can access something new and fresh. These worlds are similar to the universal thoughts often being handed out to people all over the globe.

Who can pick up this transmission? Who will take action on the brilliance that comes to you?

TWELVE

Accept the Loss of Everything Important to You, Including Your Own Life

Within the flow of time, all things ultimately are lost, and loss, like gain, is a constant. Accept the loss of everything important to you, including your own life, and you will attract what you are afraid of losing – because gain is also a constant.

Time implies loss, because it implies change. If nothing changed, we would not have a way to measure or observe the passing of time. Because things change, it is inevitable that anything we value and find important will in the end be lost, including life itself.

Your willingness to accept the eventual loss of everything you care about is your only access to being truly fulfilled and happy. Holding on and guarding against losing your love, your job, your money or your life, prevents you from fully living now.

This is not about taking risks. It is about accepting the nature of life and the nature of reality itself, which includes constant loss and constant gain.

Clinging to the things you don't want to lose demonstrates your fundamental disorientation to the way life actually works. What is that? Life is a wave through which molecules of water are always moving, with some coming in and others going out. The thing to understand is that loss is always balanced by gain. There is always something new coming in to replace the loss of what has gone out.

Clinging to what you think you can't bear to lose shows that you don't trust yourself, that you don't trust life itself. But what if you have a string of failures and losses that have shaken you to the core, and you find it emotionally difficult to let go?

If that is the case, the solution is forgiveness, both of yourself and of others. Love and accept yourself. Very soon you will have restored trust in yourself and in life, now have confidence that you don't need to grasp onto that which you fear losing. Once you do that, you will be a magnet attracting what you once feared losing. Because, although it is always being lost, it is also always being found.

THIRTEEN

Pushing Through Resistance is Necessary to Achieving Success

When you have made a commitment to achieving something, resistance appears. It will appear internally as well as externally, taunting you to give up. You are faced with where you are and where you need to go. When that happens, the next step is to accept and overcome the resistance.

Are you committed to succeeding in your endeavor? Then burn the bridges, there is no going back! Many times you will think the endeavor is too much, or not worth it. You will question why you ever started, including offering excuses to yourself or another. You will say things like, "I didn't really mean it." All of that is just head trash.

Once the journey begins, the worst thing you can do is stop.

The second worst thing you can do is to change your goal and chase something else. Occasionally, a new opportunity emerges on a journey that warrants the suspending of one journey and the pursuing of another journey. Look though; is it a new idea with another big dream attached? That voice in your ear is a lie, a force more like the sirens from The Odyssey. Remember that no matter what obstacles, temptations or distractions Odysseus faced, his goal of returning home never wavered.

Finish what you start and build on that success to pursue what is next. Otherwise kill the endeavor and channel your energy into what you will now actually complete.

Always bring yourself back to the "Why" you are on the journey, and what is the outcome or destination. Keep your vision on that and push through the resistance. Resistance is just a normal and natural part of any creative endeavor worth pursuing. Get used to it, get comfortable that this is just part of it. This process of creation closes the gap between where you are and where you want to be.

FOURTEEN

The Key to Getting Unstuck in Life is to Stop Everything

At some point while pursuing your purpose you will hit a plateau or encounter a recurring problem. Though sometimes life calls for pressing forward and taking new action, you need to be able to recognize if you are "stuck". If you're truly stuck, you must take a step back and get some perspective by bringing wonder to the situation. Ancient teachers and wise sages find the answers by bringing a new perspective to the situation through a trusted advisor or mentor, not someone who just gives you answers or steps, but someone that can bring something new to the problem.

Just as a dolphin caught in a net, the dolphin needs someone else to set him loose or like a car stuck in the mud, needs a tow to get out, you need the fresh perspective of a mentor. The thinking that got you to where you are is insufficient to get you to the next level.

If you are hitting a plateau it's because where you are

today is a result of the actions you've been taking and the actions you've been taking no longer work for you at your current level. To get you to the next level you must take new actions, maybe actions you've never taken before. What actions to take is a blind spot that only a mentor can help you distinguish.

Another aspect of being stuck is if you keep hitting the same barrier or you are experiencing a recurring problem. Unless you have the patience of Thomas Edison for 10,000 attempts, it makes more sense to find out about the problem by speaking to someone who has dealt with a similar problem as you and find out how they moved through it. A mentor can help give you new insight when looking at a problem you are not able to deal with powerfully.

The key is not to look for the answers out there, rather look to solve the problem by looking inside of you. The opportunity is for something in you to be healed and transformed such that this recurring situation disappears. Ask yourself: What is there to heal or evolve within me? What is it in me that has this problem come up again and again?

But watch out, do not try to fix the problem. Trying to fix the problem will further exacerbate it. As said earlier, the key is to bring wonder to the very fact that you're stuck again and that this situation persists. A new kind of curiosity creates an opening that sheds light on a hidden part of you that has previously been concealed. By shedding light onto an internal issue that has been concealed from your view, you can resolve the internal problem, allowing the external problem to naturally transform.

FIFTEEN

A Sense of Wonder Often Comes from Nature and Being in Immediate Contact with It

The digital world gives us enormous power, but by insulating us from nature it may prevent us from feeling that sense of wonder that is the key to both happiness and achievement. The answer is not to disconnect, but to find that sense of wonder even while wired in, from the digital world itself, which is part of the amazing fact that anything happens at all.

I hiked hundreds of miles on mountainous trails the summer after completing college. It gave me a profound sense of wonder and connection to nature. The same thing happened more recently in my life, more vivid in my memory as I approach its end, with hikes around lakes, with being at the ocean, while watching a beautiful sunset, when seeing a smiling baby, while remembering being with a new lover.

That sense of appreciation gets us to a place of realizing that there is something behind each moment. You feel the ineffable, that inexpressible source of life giving birth to the moment. It points to something much more profound and even sacred, and to the ultimate mystery that anything exists at all.

We are now "always on," constantly interacting with the global world through a device in our hands. This is an amazing amount of power. We are connected to everyone who wants to be connected with us, communicating instantly.

But maybe the digital numbs our sense of wonder? How do we now consciously cultivate it, to bring back the key element that makes us human? Is it possible to achieve the same natural sense of awe while remaining wired into the world?

Though I have only begun to experience this for myself, I see no reason why not. This global brain we have made is just as much a part of the Great Mystery as a sunset or a mountain lake, a lover's smile or a baby's cry. The real difference comes, as always, from inside us, not outside us.

I invite you to help us learn to use the device in a way, even program the network of devices so it cultivates the sense of ineffable in each moment of living. How can technology keep us connected with who we are, enabling us to remember that there is something significant behind each moment we face together? What if we had the vision to design it into the digital experience?

JULIAN BERGQUIST

SIXTEEN
Embrace Your Individuality

There is no need to justify yourself to anyone (including yourself!). To justify yourself is to make yourself small.

Enjoy the discovery process and getting to know who you are. That is one of the values of this time. Don't grow up too quickly. Go faster, though, and push through to discover your preferences and the qualities that make you who you are. There is a time for this, and that is now. Make room for it.

Do you prefer guys or girls? Do you seek one career over another? Do you desire to live in a new city? Yes, these choices, do frame many other choices and opportunities. Though these things are not what matter most.

And I can assure you that soon you will discover what really matters. You'll soon want to contribute to something bigger than you. You'll attract business, life and soul partners who you can build something with. You have a gift for enchanting people into your thoughts and ideas. For what cause do you enchant? For what purpose do you suck them in? Is it for personal expression and discovery? Or are your charms offered up to a bigger cause? It is fine either way. And you can connect with something great, to give your life to something significant, something that makes a difference, that leverages your gifts and talents. Something that the world is waiting for because you are the only person alive who can give it. You don't have to search to find this. It is right here and right now.

You can be happy whichever path you choose, however this goes. Only you can make you happy. Once again, bring a sense of wonder to the significance you've given to your preferences, to these little things you have made into big things. This will all work out just fine. Standing a few years in the future, which of these things you're facing now will you remember? Put your attention on those things.

You need to be whole in your own being, be connected with who you are, and what constitutes you as a being. What constitutes you as a human separate from your family or culture? That includes the pretty and the ugly. What makes up your wounds and triggers? Once you come to accept who you are and the way you are, no longer seeking to change any of it, you can be truly whole as a person. You now stand in the world as an individual, contributing uncluttered wholeness to the world, in your teams and groups, as a full contributor.

Embrace your individuality. Don't justify anything about who you are to anybody. It just is what it is.

JULIAN BERGQUIST

SEVENTEEN

The Key to Great Relationships is Focusing on What People Care About Rather than What You Want to Say

When listening to someone else, your thoughts about them form a barrier to understanding where they are coming from.

Paying attention to what people really care about is the key to communication. Communication is a function of listening. Not just hearing the words, but also the NOT words. Where are their words coming from? You have to get out of the way. All your stories and thoughts about them, or about yourself get in the way of you being present. When you are fully present you will hear entire worlds behind their words. If you can enter their conversation at this level you can offer the most profound gift you can give any human being: for them to be truly understood, or even known at the most essential level of being.

Develop your capacity to listen above all else. With

JULIAN BERGQUIST

the digital stream, we don't show our truest selves. We can edit and put up a persona of our own choosing. Behind that person is a creature longing for a true connection where they are known and accepted for who they are and who they are not. Each person is fortunate to have even a couple people in the world who truly know them. What if you can be one of those people for the thousands of people you will interact with this year? And what if you show people how to truly listen?

Once again, bring wonderment to every moment. Curiosity to who each person is, what has them be the way they are, without judgment.

Just be compassionate for it all.

54

EPILOGUE

The prophet breathed his last and passed on. We now try to make sense of the wisdom he offered us and what to do about it. I'm still boggled by it.

In 2014, the future of big data and technology predicting what we may each desire to have or experience next is now a reality. Technology promises to be our biggest ally, providing each person a convenient lifestyle and fulfilling our preferences for just about everything – while having it all handed to us on a platter. Digital technology will make the quality of our lives easier and abundantly better.

But this future is currently bankrupt. How often does a leader with a disruptive idea or technology comprehend the ramifications of what they are bringing forth? Are they providing just another choice in the market, or are they fulfilling a deeper human need? Or are they just scratching an itch?

Are you a founder of a company or organization who is looking to build a disruptive technology or social movement that addresses the deeper human need and supports people in experiencing what matters most for them?

If so, I would like to learn about what you are doing and discover the synergy among kindred spirits.

All the Best,

Co-Founder

VITALIZATION

Contact me here:

http://www.linkedin.com/in/julianbergquist

I have given my life to helping people be powerful while facing life's biggest challenges and channel their passion and energy into what matters most.

I hope that you have found value in this short fable. I've created some additional exercises you can download to support you in a specific area such as:

1. Getting unstuck
2. Choosing from several options
3. Launching a big idea
4. Getting more results from your team

Please go to http://www.julianbergquist.com and use them to choose the right path amidst constant change, endless choice, and digital living.

JULIAN BERGQUIST

ABOUT THE AUTHOR

Julian Bergquist is the co-founder of Vitalization, a communication company that enables the founders of growing movements to make a wider impact and generate greater profits without sabotaging their current success.

Julian connects founding CEOs with their deeper aspirations and goals for their business, and supports them in activating growth strategies that fulfill on those intentions.

The full execution and alignment of the business around these strategies results in wider market penetration, increased profitability, and exit strategies that produce wealth for everyone involved.